OpenStack
-Ansibleで作る
HA環境構築手順書 [Kilo版]

日本仮想化技術株式会社=著

構成管理ツールAnsibleを活用して
実用上不可欠なHigh Availability構成を構築
<Ubuntu Server 14.04 ベース>

インプレス

- 本書は、日本仮想化技術が運営するEnterpriseCloud.jpで提供している「OpenStack構築手順書」をオンデマンド書籍として再編集したものです。
- 本書の内容は、執筆時点までの情報を基に執筆されています。紹介したWebサイトやアプリケーション、サービスは変更される可能性があります。
- 本書の内容によって生じる、直接または間接被害について、著者ならびに弊社では、一切の責任を負いかねます。
- 本書中の会社名、製品名、サービス名などは、一般に各社の登録商標、または商標です。なお、本書では©、®、TMは明記していません。

はじめに

OpenStack-Ansible は、Ansible を使用して OpenStack の構築・機能追加・アップグレードを容易にデプロイできることをめざす OpenStack Foundation 公式のプロジェクトです。

- OpenStack-Ansible
  https://github.com/openstack/openstack-ansible

本書では、2 台のサーバーを使って OpenStack-Ansible に必要な環境を準備し、実際に OpenStack をデプロイする手順を解説します。

参考文献

- OpenStack-Ansible Installation Guide
  http://docs.openstack.org/developer/openstack-ansible/install-guide/index.html

- OpenStack Cloud Computing Cookbook Third Edition
  Kevin Jackson; Cody Bunch; Egle Sigler 著
  Packt Publishing Limited.

目　次

はじめに iii

参考文献 iii

第 1 章　構築する環境について 1
1.1　サーバーの構成 1
1.2　環境構築に使用する OS 2
1.3　ネットワーク設計 3
1.4　各サーバーのネットワーク設定 5
1.5　Ubuntu Server のインストールと初期設定 6
1.6　Ubuntu Server へのログインと root 権限 8

第 2 章　OpenStack-Ansible インストール前の設定 9
2.1　パッケージインストール 9
2.2　カーネルのアップグレード 10
2.3　ネットワークデバイスの設定 10
2.4　物理ボリュームの設定 12

第 3 章　OpenStack-Ansible の設定 15
3.1　OpenStack-Ansible のダウンロード 15
3.2　openstack_user_config.yml ファイルの作成 16
3.3　user_variables.yml ファイルの設定 20

3.4	user_secrets.yml ファイルの設定	21

第 4 章　OpenStack のデプロイ　23

4.1	OpenStack-Ansible の事前準備	23
4.2	setup-hosts.yml の実行	24
4.3	haproxy-install.yml の実行	24
4.4	setup-infrastructure.yml の実行	25
4.5	setup-openstack.yml の実行	25

第 5 章　コンテナの操作　27

5.1	コンテナ操作コマンド一覧	27
5.2	コンテナ操作の出力例	27

第 6 章　OpenStack の操作　31

6.1	プロジェクト（テナント）作成	31
6.2	ユーザー作成	32
6.3	ロール作成とユーザー割り当て	32
6.4	Glance へのイメージ登録	33
6.5	仮想ネットワーク設定	34
6.6	セキュリティグループの設定変更	39
6.7	インスタンスの起動	40
6.8	FloatingIP の設定	43
6.9	Cinder の設定	44
6.10	インスタンスの動作確認	45

第 7 章　Tips とトラブルシューティング　49

7.1	コンテナを削除し再構築する	49
7.2	OpenStack 構築後 Controller サーバーを再起動すると正常に動作しない	50
7.3	ネットワークの冗長化	50
7.4	YAML ファイルの構文について	50

付録 A　FAQ フォーラム参加特典について　53

第1章 構築する環境について

1.1 サーバーの構成

物理サーバー

本書は OpenStack 環境を Controller サーバー 1 台、Compute サーバー 1 台の計 2 台を物理サーバー上に構築することを想定しています。最低限必要となるスペックは以下のとおりです。

	Controller サーバー	Compute サーバー
CPU	8 コア以上	2 コア
メモリ	16GB 以上	8GB 以上
ディスク1	80GB	40GB
ディスク2	任意のサイズ	(不要)
NIC	2 つ以上	2 つ以上

LXC コンテナの構成

OpenStack-Ansible ではコンポーネント毎に LXC コンテナが作成され、そこへ各コンポーネントをインストールされます。

重要なコンポーネントについては可用性を高めるために複数のコンテナを立てクラスタを構成します。クラスタのノード数は、データの書き込みが行われるコンポーネントは奇数台（3 台以上）が望ましい構成となります。これは、障害によりデータ競合が発生した場合多数決で処理されるためです。

今回は検証環境の構築を想定しているため 1 台の物理サーバーの中で組まれていますが、本番運用で構築する場合は Controller サーバーを 3 台以上用意することを推奨します。

Controller サーバーに作成される LXC コンテナは以下のとおりです。

コンテナ名	分類	用途
galera_container-01	Infrastructure	Galera クラスタ（DB サービス）
galera_container-02	Infrastructure	Galera クラスタ（DB サービス）
galera_container-03	Infrastructure	Galera クラスタ（DB サービス）
rabbit**mq**container-01	Infrastructure	RabbitMQ（メッセージキューサービス）
rabbit**mq**container-02	Infrastructure	RabbitMQ（メッセージキューサービス）
rabbit**mq**container-03	Infrastructure	RabbitMQ（メッセージキューサービス）
memcached_container	Infrastructure	memcacheed
repo_container-01	Infrastructure	コンテナ向けローカルリポジトリ
repo_container-02	Infrastructure	コンテナ向けローカルリポジトリ
rsyslog_container	Infrastructure	Syslog サービス
utility_container	Infrastructure	OpenStack CUI クライアントツール
keystone_container-01	OpenStack	Keystone サービス
keystone_container-02	OpenStack	Keystone サービス
glance_container	OpenStack	Glance サービス
cinder**api**container	OpenStack	Cinder API サービス
cinder**scheduler**container	OpenStack	Cinder スケジューラー
nova**api**metadata_container	OpenStack	Nova Metadata API サービス
nova**api**os**compute**container	OpenStack	Nova Compute API サービス
nova**cert**container	OpenStack	Nova Cert サービス（証明書管理）
nova**conductor**container	OpenStack	Nova Conductor サービス
nova**console**container	OpenStack	Nova Console サービス
nova**scheduler**container	OpenStack	Nova Scheduler サービス
neutron**agents**container	OpenStack	Neutron agent サービス
neutron**server**container	OpenStack	Neutron server(API) サービス
heat**api**scontainer	OpenStack	Heat API サービス
heat**engine**container	OpenStack	Heat Engine サービス
horizon_container-01	OpenStack	Horizon(Dashboard) サービス
horizon_container-02	OpenStack	Horizon(Dashboard) サービス

1.2 環境構築に使用するOS

OS は Ubuntu Server を使用します。Ubuntu Server では新しいハードウェアのサポートを積極的に行うディストリビューションです。そのため、Linux Kernel のバージョンを Trusty の場合は 14.04.2 以降の LTS のポイントリリースごとに、スタンダート版の Ubuntu と同様のバージョンに置き換えてリリースしています。

- https://wiki.ubuntu.com/Kernel/LTSEnablementStack

一般的な利用では特に問題ありませんが、OpenStack と SDN のソリューションを連携した環境を作る場合などに、Linux Kernel や OS のバージョンを考慮しなくてはならない場合があります。また、Trusty の場合は Linux Kernel v3.13 以外のバージョンのサポート期間は Trusty のサポート期間と同様ではありません。期限が切れた後は Linux Kernel v3.13 にダウングレードするか、新しい Linux Kernel をインストールすることができるメタパッケージを手動で導入する必要がありますので注意してください。

本書ではサポート期間が長く、Trusty の初期リリースの標準カーネルである Linux Kernel v3.13 を使うために、以下の URL よりイメージをダウンロードした Ubuntu Server 14.04.1 LTS(以下 Ubuntu Server) のイメージを使ってインストールします。インストール後 `apt-get dist-upgrade` を行って最新のアップデートを適用した状態にしてください。Trusty ではこのコマンドを実行してもカーネルのバージョンがアップグレードされることはありません。

本書は 3.13.0-68 以降のバージョンのカーネルで動作する Ubuntu 14.04.1 を想定しています。

- http://old-releases.ubuntu.com/releases/14.04.1/ubuntu-14.04.1-server-amd64.iso

1.3 ネットワーク設計

今回の例では以下に述べるネットワーク構成で構築する前提で解説します。

物理構成

双方のサーバーの em1 には管理用、em2 にはサービス用のトラフィックが流れるように構築します。それぞれのインターフェースは VLAN を使用して複数のセグメントを扱うため、スイッチを経由する場合は**タグ VLAN を有効（Trunk モード）**にする必要があります。

ホスト管理用ネットワーク

管理用イントラネットワークに接続するネットワークセグメントで、OS インストール時に設定します。apt-get 等を実行するときこのネットワークを経由するためインターネットへアクセスできる環境である必要があります。今回は物理インターフェース em1 へ割り当てます。

- 物理インターフェース：em1
- VLAN：Untagged/Native
- ブリッジインターフェース：なし

第 1 章 構築する環境について

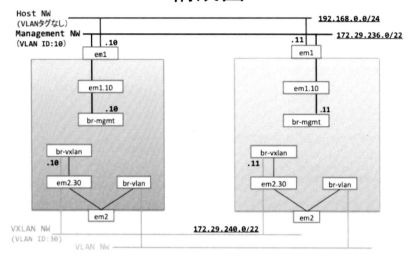

図 1.1 ネットワーク構成図

- ネットワークセグメント：192.168.0.0/24
- ゲートウェイ：192.168.0.1
- DNS サーバー：8.8.8.8

コンテナ管理用ネットワーク

　OpenStack-Ansible 環境の各コンテナ間で使用する独立したネットワークセグメントです。各サービス間の API 連携はこのネットワークを使用して通信されます。今回はタグ VLAN を使用して物理インターフェース em1 へ割り当てます。

- 物理インターフェース：em1
- VLAN ID：10
- ブリッジインターフェース：br-mgmt
- ネットワークセグメント：172.29.236.0/22

プロジェクト（テナント）用 VLAN ネットワーク

OpenStack 構築後、各プロジェクト（テナント）のサービス用に割り当てるネットワークとして使用します。このネットワークではタグ VLAN を使用して複数のセグメントを割り当てられるようにします。インターネットへ接続する必要がある場合は別途接続先のスイッチ等で VLAN の設定を行ってください。今回は物理インターフェース em2 を使用します。

- 物理インターフェース：em2
- VLAN ID：任意（OpenStack 構築後ネットワーク作成時に指定）
- ブリッジインターフェース：br-vlan

プロジェクト（テナント）用 VXLAN ネットワーク

OpenStack 構築後、各プロジェクト（テナント）の内部用に割り当てるネットワークとして使用します。このネットワークでは VXLAN を使用してプロジェクト内で作成されたネットワークをトンネル化します。今回はタグ VLAN を使用して物理インターフェース em2 へ割り当てます。

- 物理インターフェース：em2
- VLAN ID：30
- ブリッジインターフェース：br-vxlan
- ネットワークセグメント：172.29.240.0/22

1.4 各サーバーのネットワーク設定

各サーバーの設定するアドレスは以下の通りです。

Controller サーバー

インターフェース	em1	em1.10	em2	em2.30
IP アドレス	192.168.0.10	172.29.236.10	（なし）	172.29.240.10
ネットマスク	255.255.255.0	255.255.252.0	（なし）	255.255.252.0
ゲートウェイ	192.168.0.1	-	-	-
ネームサーバー	8.8.8.8	-	-	-

Compute サーバー

インターフェース	em1	em1.10	em2	em2.30
IP アドレス	192.168.0.11	172.29.236.11	(なし)	172.29.240.11
ネットマスク	255.255.255.0	255.255.252.0	(なし)	255.255.252.0
ゲートウェイ	192.168.0.1	-	-	-
ネームサーバー	8.8.8.8	-	-	-

1.5 Ubuntu Server のインストールと初期設定

Ubuntu Server のインストール

2 台のサーバーに対し、Ubuntu Server をインストールします。要点は以下の通りです。

- 優先ネットワークインターフェースを em1 に指定

- インターネットへ接続するインターフェースは em1 を使用するため、インストール中は em1 を優先ネットワークとして指定します。
- OS は最小インストール
- パッケージ選択では OpenSSH server のみ選択

インストール時の設定パラメータ例

設定項目	設定例
初期起動時の Language	English
起動	Install Ubuntu Server
言語	English - English
地域の設定	other → Asia → Japan
地域の言語	United States - en_US.UTF-8
キーボードレイアウトの認識	No
キーボードの言語	Japanese → Japanese
優先する NIC	em1: Ethernet
ホスト名	それぞれのサーバー名 (controller, compute)
ユーザ名とパスワード	フルネームで入力
アカウント名	ユーザ名のファーストネームで設定される
パスワード	任意のパスワード
Weak password（出ない場合も）	Yes を選択
ホームの暗号化	任意
タイムゾーン	Asia/Tokyo であることを確認
パーティション設定	Guided - use entire disk and set up LVM
パーティション選択	sda を選択
パーティション書き込み	Yes を選択
パーティションサイズ	デフォルトのまま
変更の書き込み	Yes を選択
HTTP proxy	環境に合わせて任意
アップグレード	No automatic updates を推奨
ソフトウェア	OpenSSH server のみ選択
GRUB	Yes を選択
インストール完了	Continue を選択

筆者注：
Ubuntu インストール時に選択した言語がインストール後も使われます。
Ubuntu Server で日本語の言語を設定した場合、標準出力や標準エラー出力が文字化けしたり、作成されるキーペア名が文字化けするなど様々な問題が起きますので、言語は英語を設定されることを推奨します。

プロキシーの設定

外部ネットワークとの接続にプロキシーの設定が必要な場合は、apt コマンドを使ってパッケージの照会やダウンロードを行うために次のような設定をする必要があります。

- システムのプロキシー設定

```
# vi /etc/environment
http_proxy="http://proxy.example.com:8080/"
https_proxy="https://proxy.example.com:8080/"
```

- APT のプロキシー設定

```
# vi /etc/apt/apt.conf
Acquire::http::proxy "http://proxy.example.com:8080/";
Acquire::https::proxy "https://proxy.example.com:8080/";
```

より詳細な情報は下記のサイトの情報を確認ください。

- https://help.ubuntu.com/community/AptGet/Howto

1.6 Ubuntu Serverへのログインとroot権限

Ubuntu はデフォルト設定で root ユーザーの利用を許可していないため、root 権限が必要となる作業は以下のように行ってください。

- root ユーザーで直接ログインできないので、インストール時に作成したアカウントでログインする。
- root 権限が必要な場合には、sudo コマンドを使用する。
- root で連続して作業したい場合には、sudo -i コマンドでシェルを起動する。

第2章 OpenStack-Ansible インストール前の設定

本章では、OpenStack パッケージのインストール前に各々のサーバーで以下の設定を行います。

- パッケージ インストール
- カーネルのアップグレード
- ネットワークデバイスの設定
- 物理ボリュームの設定

2.1 パッケージインストール

OpenStack-Ansible 環境を構築するにあたり、必要となるパッケージのインストールを行います。

Controller サーバー

```
controller$ sudo -i
controller# apt-get update
controoler$ apt-get install aptitude build-essential git python-dev bridge-utils debootstrap ifenslave ifenslave-2.6 lsof lvm2 ntp ntpdate openssh-server sudo tcpdump vlan python-crypto python-yaml
```

Compute サーバー

```
compute$ sudo -i
compute# apt-get update
compute# apt-get install bridge-utils debootstrap ifenslave ifenslave-2.6 lsof lvm2
ntp ntpdate openssh-server sudo tcpdump vlan
```

タグ VLAN の有効化

VLAN 有効化するには、vlan パッケージインストール後モジュールの有効化をする必要があります。双方のサーバーでタグ VLAN を有効化します。

```
# echo '8021q' >> /etc/modules
```

2.2 カーネルのアップグレード

OpenStack-Ansible では Linux カーネルのバージョンが 3.13.0-34 以上の環境を要求します。Controller サーバーと Compute サーバーの両方でカーネルのアップグレード行います。

```
# apt-get dist-upgrade
```

2.3 ネットワークデバイスの設定

Controller サーバーと Compute サーバーのネットワークを設定します。

Controller サーバーの IP アドレスの設定

```
controller:$ vi /etc/network/interfaces

auto em1
iface em1 inet static
    address 192.168.0.10
    netmask 255.255.255.0
    gateway 192.168.0.1
    dns-nameservers 8.8.8.8

iface em1.10 inet manual
    vlan-raw-device em1

auto em2
iface em2 inet manual
```

```
iface em2.30 inet manual
  vlan-raw-device em2

auto br-mgmt
iface br-mgmt inet static
  bridge_stp off
  bridge_waitport 0
  bridge_fd 0
  bridge_ports em1.10
  address 172.29.236.10
  netmask 255.255.252.0

auto br-vlan
iface br-vlan inet manual
  bridge_stp off
  bridge_waitport 0
  bridge_fd 0
  bridge_ports em2

auto br-vxlan
iface br-vxlan inet static
  bridge_stp off
  bridge_waitport 0
  bridge_fd 0
  bridge_ports em2.30
  address 172.29.240.10
  netmask 255.255.252.0
```

ComputeサーバーのIPアドレスの設定

```
ubuntu@compute:~$ vi /etc/network/interfaces

auto em1
iface em1 inet static
  address 192.168.0.11
  netmask 255.255.255.0
  gateway 192.168.0.1
  dns-nameservers 8.8.8.8

iface em1.10 inet manual
  vlan-raw-device em1

auto em2
iface em2 inet manual

iface em2.30 inet manual
  vlan-raw-device em2

auto br-mgmt
iface br-mgmt inet static
  bridge_stp off
```

```
  bridge_waitport 0
  bridge_fd 0
  bridge_ports em1.10
  address 172.29.236.11
  netmask 255.255.252.0

auto br-vlan
iface br-vlan inet manual
  bridge_stp off
  bridge_waitport 0
  bridge_fd 0
  bridge_ports em2

auto br-vxlan
iface br-vxlan inet static
  bridge_stp off
  bridge_waitport 0
  bridge_fd 0
  bridge_ports em2.30
  address 172.29.240.11
  netmask 255.255.252.0
```

ホストの再起動

設定後、Controller サーバーと Compute サーバーを再起動します。

```
# shutdown -r now
```

2.4 物理ボリュームの設定

Controller サーバーに Cinder 用の LVM ボリュームグループを作成します。

まず、デバイス /dev/sdb に対し LVM 物理ボリュームを作成します。

```
controller$ sudo pvcreate /dev/sdb
```

LVM 物理ボリュームが作成されたことを確認します。

```
controller$ sudo pvscan
  PV /dev/sdb                        lvm2 [279.37 GiB]
  Total: 1 [279.37 GiB] / in use: 0 [0    ] / in no VG: 1 [279.37 GiB]
```

続いて、LVM ボリュームグループを作成し先ほど作成した物理ボリュームを登録します。

```
controller$ sudo vgcreate cinder-volumes /dev/sdb
```

ボリュームグループが作成されたか確認します。

2.4 物理ボリュームの設定

```
controller$ sudo vgscan
  Reading all physical volumes.  This may take a while...
  Found volume group "cinder-volumes" using metadata type lvm2
```

第3章 OpenStack-Ansible の設定

本章では、OpenStack-Ansible を GitHub からダウンロードし、設定ファイル（YAML 形式）を編集します。

設定に用いる対象ファイルは以下の3ファイルです。

- openstack**user**config.yml
- user_variables.yml
- user_sercret.yml

3.1 OpenStack-Ansibleのダウンロード

GitHub から OpenStack-Ansible をダウンロードし、/opt/openstack-ansible ディレクトリへ格納します。

ダウンロード

```
controller$ sudo git clone https://github.com/openstack/openstack-ansible /opt/openstack-ansible
```

バージョンの指定

使用する OpenStack-Ansible のバージョンのタグをチェックアウトします。（今回は 11.2.6 を指定します。）

```
controller$ cd /opt/openstack-ansible/
controller$ git tag -l
controller$ sudo git checkout 11.2.6
```

チェックアウトが正しく行われたか確認します。

```
controller$ git branch -a
```

(detached from 11.2.6) と表示されていることを確認します。

設定ファイルのコピー

設定ファイルを /etc/openstack_deploy へコピーします。

```
controller$ sudo cp -a /opt/openstack-ansible/etc/openstack_deploy
/etc/openstack_deploy
```

3.2　openstack_user_config.yml ファイルの作成

openstack_user_config.yml ファイルを作成し、以下の設定を記述していきます。

```
controller$ sudo vi /etc/openstack_deploy/openstack_user_config.yml
```

cidr_networks セクション

OpenStack-Ansible で使用するネットワークを定義します。今回の環境では以下 2 つのネットワークを設定します。

- management：コンテナ用ネットワーク
- tunnel：プロジェクト（テナント）用ネットワーク

```
---           ← ファイルの先頭行に YAML のセパレーション記号「---」を記述
cidr_networks:
  management: 172.29.236.0/22
  tunnel: 172.29.240.0/22
```

used_ips セクション

OpenStack-Ansible では、LXC コンテナで使用する IP アドレスを定義したアドレス範囲内

3.2 openstack_user_config.yml ファイルの作成

で自動生成します。すでに使用しているアドレスや使用されたくないアドレスがある場合は、user_ips セクションで予約します。

アドレスの指定は単一もしくは範囲で指定します。範囲で使用する場合はカンマ区切りで下限と上限の IP アドレスを指定します。

```
used_ips:
  - 172.29.236.1,172.29.236.31
  - 172.29.240.1,172.29.240.31
```

global_overrides セクション

OpenStack-Ansible では HAProxy を使用してクラスタ構成しているサービスのロードバランシングを行います。設定内容は以下のとおりです。

項目	内容
internal_lb_vip_address	内部向けアドレス。コンテナ管理用ネットワークセグメント内のアドレスを指定
external_lb_vip_address	外部向けアドレス。ホスト管理用ネットワークセグメント内のアドレスを指定
lb_name	ロードバランサーの名前
tunnel_bridge	VXLAN で使用するブリッジを指定
management_bridge	管理用ネットワーク使用するブリッジを指定
probider_networks	各ネットワークの詳細を設定

```
global_overrides:
  internal_lb_vip_address: 172.29.236.10
  external_lb_vip_address: 192.168.0.10
  lb_name: haproxy
  tunnel_bridge: "br-vxlan"
  management_bridge: "br-mgmt"
  provider_networks:
    - network:
        group_binds:
          - all_containers
          - hosts
        type: "raw"
        container_bridge: "br-mgmt"
        container_interface: "eth1"
        container_type: "veth"
        ip_from_q: "management"
        is_container_address: true
        is_ssh_address: true
    - network:
        group_binds:
          - neutron_linuxbridge_agent
        container_bridge: "br-vxlan"
        container_type: "veth"
```

第 3 章　OpenStack-Ansible の設定

```
        container_interface: "eth10"
        ip_from_q: "tunnel"
        type: "vxlan"
        range: "1:1000"
        net_name: "vxlan"
    - network:
        group_binds:
          - neutron_linuxbridge_agent
        container_bridge: "br-vlan"
        container_type: "veth"
        container_interface: "eth11"
        type: "vlan"
        range: "1:1"
        net_name: "vlan"
```

shared-infra_hosts セクション

　OpenStack の各コンポーネントと連携するサービス（GaleraDB クラスタ、RabbitMQ、Memcached 等）を稼働させるホストを指定します。

　今回はコントローラサーバー（controller）上に GaleraDB クラスタ用のコンテナを 3 台、RabbitMQ 用のコンテナを 3 台、Memcached 用のコンテナを 1 台インストールする設定とします。

```
shared-infra_hosts:
  controller-01:
    affinity:
      galera_container: 3
      rabbit_mq_container: 3
    ip: 172.29.236.10
```

　なお、物理マシンを複数台用意する場合は以下のように指定します。

```
shared-infra_hosts:
  infra-01:
    ip: 172.29.236.21
  infra-02:
    ip: 172.29.236.22
  infra-03:
    ip: 172.29.236.23
```

os-infra_hosts セクション

　OpenStack の各コンポーネントを稼働させるホストを指定します。今回はコントローラサーバー（controller）上に Glance API 用のコンテナを 1 台、Nova API 用のコンテナを 1 台、Horizon 用のコンテナを 2 台インストールする設定とします。

```
os-infra_hosts:
  controller-01:
    affinity:
      horizon_container: 2
    ip: 172.29.236.10
```

storage-infra_host セクション

　Cinder API を稼働させるホストを指定します。今回はコントローラサーバー（controller）上に Cinder API 用のコンテナを 1 台インストールする設定とします。

```
storage-infra_hosts:
  controller-01:
    ip: 172.29.236.10
```

identity_hosts セクション

　Keystone を稼働させるホストを指定します。今回はコントローラサーバー（controller）上に Keystone API 用のコンテナを 2 台インストールする設定とします。

```
identity_hosts:
  controller-01:
    affinity:
      keystone_container: 2
    ip: 172.29.236.10
```

compute_hosts セクション

　Nova compute を稼働させるホストを指定します。今回はコンピュートサーバー（compute）に Nova compute サービスをデプロイする設定とします。

```
compute_hosts:
  compute-01:
    ip: 172.29.236.11
```

storage_hosts セクション

　Cinder ストレージサーバーを稼働させるホストを指定とします。今回はコントローラーサーバーに Cinder volume サービスをデプロイし、ボリュームを LVM ボリュームグループ上に作成します。

```
storage_hosts:
  controller-01:
    ip: 172.29.236.10
    container_vars:
      cinder_backends:
        limit_container_types: cinder_volume
        lvm:
          volume_backend_name: LVM_iSCSI
          volume_driver: cinder.volume.drivers.lvm.LVMVolumeDriver
          volume_group: cinder-volumes
```

network_hosts セクション

Neutron API サービスを稼働させるホストを指定します。今回は Controller サーバーに Neutron サービスをデプロイする設定とします。

```
network_hosts:
  controller-01:
    ip: 172.29.236.10
```

repo-infra_hosts セクション

パッケージレポジトリを稼働させるホストを指定します。今回は Controller サーバーにパッケージレポジトリ用のコンテナを 2 台インストールする設定とします。

```
repo-infra_hosts:
  controller-01:
    affinity:
      repo_container: 2
    ip: 172.29.236.10
```

haproxy_hosts セクション

HAProxy を稼働させるホストを指定します。今回は Controller サーバーに HAProxy をインストールする設定とします。

```
haproxy_hosts:
  controller-01:
    ip: 172.29.236.10
```

3.3　user_variables.yml ファイルの設定

本ファイルにおいては OpenStack コンポーネント（Glance、Nova、Apache 等）のオプショ

ン設定を行います。以下のコマンドで設定ファイルを編集します。

```
controller$ sudo vi /etc/openstack_deploy/user_variables.yml

## Glance Options
# Set glance_default_store to "swift" if using Cloud Files or swift backend
# or "rbd" if using ceph backend; the latter will trigger ceph to get
# installed on glance
glance_default_store: file          ← コメントが外れていることを確認
(省略)
# This defaults to KVM, if you are deploying on a host that is not KVM capable
# change this to your hypervisor type: IE "qemu", "lxc".
nova_virt_type: kvm                 ← コメントを外す
nova_cpu_allocation_ratio: 2.0      ← コメントを外す
nova_ram_allocation_ratio: 1.0      ← コメントを外す
```

3.4　user_secrets.ymlファイルの設定

　user_secrets.yml ファイルでは、OpenStack 内の各コンポーネントで使用する初期パスワードをまとめています。以下の操作を行い、user_secrets.yml ファイルにパスワードを生成します。

```
controller$ cd /opt/openstack-ansible/
controller$ sudo scripts/pw-token-gen.py --file
/etc/openstack_deploy/user_secrets.yml
```

　生成したパスワードを確認します。

```
controller$ sudo less /etc/openstack_deploy/user_secrets.yml
```

　上記ファイルのなかの keystone_auth_admin_password が horizon へログインするためのパスワードとなります。パスワードを変更したい場合は、以下のように任意の文字へ変更します。

　　例）：

```
controller$ sudo vi /etc/openstack_deploy/user_secrets.yml

(省略)
keystone_auth_admin_password: password      # 任意のパスワードに変更
```

第4章 OpenStackのデプロイ

　本章では、前章で設定したYAMLファイルを元にAnsibleを利用してOpenStackをデプロイします。実行するスクリプトやAnsible Playbookは /opt/openstack-ansible/ ディレクトリに格納されており、次項の手順に従いOpenStackをインストールします。

4.1　OpenStack-Ansibleの事前準備

　以下のスクリプトを実行し、OpenStack-Ansibleの事前準備を行います。スクリプトでは必要なパッケージのインストールや設定等が実行されます。

```
controller$ cd /opt/openstack-ansible/
controller$ sudo scripts/bootstrap-ansible.sh
```

　Ansibleは公開鍵認証を使ってログインできることが前提条件となるため、以下の手順でComputeサーバーへSSH公開鍵を登録します。

　Controllerサーバーでrootアカウントの SSH 公開鍵を確認します。

```
controller$ sudo -i
controller# cat /root/.ssh/id_rsa.pub
```

　上記コマンドで表示された結果を控えておき、Computeサーバーの authorized_keys ファイルへ公開鍵を貼り付けます。

```
compute$ sudo -i
compute# mkdir /root/.ssh
compute# vi /root/.ssh/authorized_keys
```

　controllerサーバーから、SSHでrootログインできることを確認します。

```
controller# ssh localhost
The authenticity of host 'localhost (::1)' can't be established.
ECDSA key fingerprint is 29:48:13:48:22:69:68:34:e9:19:ae:1e:39:58:83:ac.
Are you sure you want to continue connecting (yes/no)? yes

controller# exit

controller# ssh 172.29.236.11
The authenticity of host 'localhost (::1)' can't be established.
ECDSA key fingerprint is 29:48:13:48:22:69:68:34:e9:19:ae:1e:39:58:83:ac.
Are you sure you want to continue connecting (yes/no)? yes
Warning: Permanently added '172.29.236.11' (ECDSA) to the list of known hosts.
Welcome to Ubuntu 14.04.3 LTS (GNU/Linux 3.13.0-76-generic x86_64)

controller# exit
```

4.2　setup-hosts.ymlの実行

　前項でスクリプトを実行すると、Playbookを実行するために使用するopenstack-ansibleと呼ばれるラッパースクリプトがインストールされます。そのスクリプトを用いてsetup-hosts.ymlを実行し、各種LXCコンテナの作成とComputeサーバーの事前準備を行います。

```
controller$ cd /opt/openstack-ansible/playbooks/
controller$ sudo openstack-ansible setup-hosts.yml -vvv
```

　実行結果がすべてOKと出力されれば問題ありません。Playbookが異常終了した場合は、原因を対処し以下のコマンドでPlaybookを再開してください。

```
controller$ sudo openstack-ansible setup-hosts.yml --limit
@/home/ubuntu/setup-hosts.retry
```

4.3　haproxy-install.ymlの実行

　haproxy-install.ymlを実行し、HAProxyのインストールを行います。

```
controller$ sudo openstack-ansible haproxy-install.yml -vvv
```

　実行結果がすべてOKと出力されれば問題ありません。Playbookが異常終了した場合は、原因を対処し以下のコマンドでPlaybookを再開してください。

```
controller$ openstack-ansible haproxy-install.yml --limit
@/home/ubuntu/haproxy-install.retry
```

4.4 setup-infrastructure.ymlの実行

setup-infrastructure.yml を実行しインフラ環境をセットアップします。Playbook を実行すると以下の環境が構築されます。

- MariaDB Galera Cluster
- RabbitMQ
- memcached
- syslogd

```
controller$ sudo openstack-ansible setup-infrastructure.yml -vvv
```

実行結果がすべて OK と出力されれば問題ありません。Playbook が異常終了した場合は、原因を対処し以下のコマンドで Playbook を再開してください。

```
controller$ sudo openstack-ansible setup-infrastructure.yml --limit
@/home/ubuntu/setup-infrastructure.retry
```

4.5 setup-openstack.ymlの実行

setup-openstack.yml を実行し、以下の OpenStack コンポーネントをセットアップします。

- Keystone
- Galnce
- Cinder
- Nova
- Neutron
- Horizon

```
controller$ sudo openstack-ansible setup-openstack.yml -vvv
```

実行結果がすべて OK と出力されれば問題ありません。Playbook が異常終了した場合は、原

第 4 章　OpenStack のデプロイ

因を対処し以下のコマンドで Playbook を再開してください。

```
controller$ sudo openstack-ansible setup-openstack.yml --limit
@/home/ubuntu/setup-openstack.retry
```

第5章 コンテナの操作

OpneStack-Ansibleで構築されたコンテナは、LXCの各コマンドを使って操作します。なお、コンテナの操作はrootユーザーで実行します。あらかじめsudoコマンドでrootユーザーに切り替えてください。

```
controller$ sudo -i
```

5.1 コンテナ操作コマンド一覧

内容	コマンド
コンテナの一覧表示	lxc-ls --fansy
コンテナの詳細表示	lxc-info --name <コンテナ名>
コンテナの起動	lxc-start --daemon --name <コンテナ名>
コンテナの停止	lxc-stop --name <コンテナ名>
コンテナ内へログイン	lxc-attach --name <コンテナ名>

5.2 コンテナ操作の出力例

ここでは、コンテナの一覧表示と、コンテナ内へのログイン方法の例を紹介します。

- コンテナの一覧表示

```
root@controller:~# lxc-ls --fancy
NAME                                                        STATE     IPV4
IPV6    AUTOSTART
```

```
--------------------------------------------------------------------------------
controller-01_cinder_api_container-3157cd2c           RUNNING  10.0.3.156,
172.29.236.56                   -        YES (onboot, openstack)
controller-01_cinder_scheduler_container-b6a20022     RUNNING  10.0.3.85,
172.29.236.243                  -        YES (onboot, openstack)
controller-01_galera_container-29d52973               RUNNING  10.0.3.157,
172.29.236.46                   -        YES (onboot, openstack)
controller-01_galera_container-2ea457da               RUNNING  10.0.3.190,
172.29.236.155                  -        YES (onboot, openstack)
controller-01_galera_container-f8883adf               RUNNING  10.0.3.113,
172.29.236.211                  -        YES (onboot, openstack)
controller-01_glance_container-62ebe7a8               RUNNING  10.0.3.155,
172.29.236.55                   -        YES (onboot, openstack)
controller-01_heat_apis_container-9507e331            RUNNING  10.0.3.47,
172.29.236.190                  -        YES (onboot, openstack)
controller-01_heat_engine_container-ee16301a          RUNNING  10.0.3.240,
172.29.236.90                   -        YES (onboot, openstack)
controller-01_horizon_container-58bc8110              RUNNING  10.0.3.36,
172.29.236.83                   -        YES (onboot, openstack)
controller-01_horizon_container-a5a061f5              RUNNING  10.0.3.98,
172.29.236.222                  -        YES (onboot, openstack)
controller-01_keystone_container-00302b5e             RUNNING  10.0.3.136,
172.29.236.9                    -        YES (onboot, openstack)
controller-01_keystone_container-9aab73a6             RUNNING  10.0.3.173,
172.29.236.129                  -        YES (onboot, openstack)
controller-01_memcached_container-44d75a70            RUNNING  10.0.3.175,
172.29.236.43                   -        YES (onboot, openstack)
controller-01_neutron_agents_container-68879ee5       RUNNING  10.0.3.154,
172.29.240.152, 172.29.236.166  -        YES (onboot, openstack)
controller-01_neutron_server_container-0cf61dc6       RUNNING  10.0.3.167,
172.29.236.113                  -        YES (onboot, openstack)
controller-01_nova_api_metadata_container-5175080c    RUNNING  10.0.3.148,
172.29.236.91                   -        YES (onboot, openstack)
controller-01_nova_api_os_compute_container-debaf1ee  RUNNING  10.0.3.180,
172.29.236.220                  -        YES (onboot, openstack)
controller-01_nova_cert_container-f08e46b0            RUNNING  10.0.3.222,
172.29.236.41                   -        YES (onboot, openstack)
controller-01_nova_conductor_container-86281c5d       RUNNING  10.0.3.82,
172.29.236.197                  -        YES (onboot, openstack)
controller-01_nova_console_container-cf6d647d         RUNNING  10.0.3.109,
172.29.236.196                  -        YES (onboot, openstack)
controller-01_nova_scheduler_container-47255f60       RUNNING  10.0.3.133,
172.29.236.143                  -        YES (onboot, openstack)
controller-01_rabbit_mq_container-4030544d            RUNNING  10.0.3.72,
172.29.236.24                   -        YES (onboot, openstack)
controller-01_rabbit_mq_container-9078f6fd            RUNNING  10.0.3.181,
172.29.236.236                  -        YES (onboot, openstack)
controller-01_rabbit_mq_container-edf993d7            RUNNING  10.0.3.184,
172.29.236.30                   -        YES (onboot, openstack)
controller-01_repo_container-90aef04e                 RUNNING  10.0.3.95,
172.29.236.217                  -        YES (onboot, openstack)
controller-01_repo_container-ee57a601                 RUNNING  10.0.3.123,
172.29.236.10                   -        YES (onboot, openstack)
```

```
controller-01_utility_container-e90a7d93              RUNNING  10.0.3.135,
172.29.236.163                    -          YES (onboot, openstack)
```

- コンテナ内へログイン

```
root@controller:~# lxc-attach --name controller-01_utility_container-e90a7d93
root@controller-01_utility_container-e90a7d93:~#    ← コンテナ内のコマンド入力待ちとなっている
```

なお、コンテナ名は以下のような命名規則でつけられます。

```
[ホスト名]_[サービス名]_container_[ランダムな文字列]
```

- [ホスト名] には openstack_user_config.yml で設定した名前が入ります。このホスト名は OS のホスト名とは異なる名前を設定することもできます。
- [ランダムな文字列] には自動的に生成された 16 進数 8 桁が入ります

第6章　OpenStackの操作

本章ではOpenStack構築後の基本的な操作手順を記載します。

各コンポーネントのクライアントコマンドはutilityコンテナに納められています。CUIでOpenStackを操作には以下のコマンドでutilityコンテナへログインしてください。

```
controller# lxc-attach --name controller_utility_container
controller_utility_container# cd /root/
```

今回は、「demoプロジェクト」と「demoユーザー」を作成し、ネットワークと動作確認用インスタンスを構築します。

6.1　プロジェクト（テナント）作成

以下コマンドでDemo Projectというプロジェクト（テナント）を作成します。

- adminユーザー環境変数ファイルの読み込み

```
controller_utility_container# source ~/openrc
```

- Demo Projectの作成

```
controller_utility_container# openstack project create --description "Demo Project" demo
+-------------+--------------------------------+
| Field       | Value                          |
+-------------+--------------------------------+
| description | Demo Project                   |
```

```
| domain_id  | default                          |
| enabled    | True                             |
| id         | 8e19b43b0403413a9f733957d41d4c99 |
| name       | demo                             |
+------------+----------------------------------+
```

6.2 ユーザー作成

以下コマンドで demo という名前のユーザーを作成します。

```
controller_utility_container# openstack user create --password-prompt demo
User Password:                                        ←パスワードを入力
Repeat User Password:                                 ←パスワードを再度入力
+------------+----------------------------------+
| Field      | Value                            |
+------------+----------------------------------+
| domain_id  | default                          |
| enabled    | True                             |
| id         | 776a28828ff44378bc5a4efa26c36fc6 |
| name       | demo                             |
+------------+----------------------------------+
```

6.3 ロール作成とユーザー割り当て

以下コマンドで user という名前のロールを作成します。

```
controller_utility_container# openstack role create user
+-------+----------------------------------+
| Field | Value                            |
+-------+----------------------------------+
| id    | 08c684ecba574b1b870a0ca487400fb3 |
| name  | user                             |
+-------+----------------------------------+
```

作成したプロジェクトにユーザーとロールを割り当てます。

```
controller_utility_container# openstack role add --project demo --user demo user
```

作成したプロジェクト向けの環境変数ファイルを作成します。

```
controller_utility_container# vi ~/demo-openrc

export OS_PROJECT_DOMAIN_ID=default
export OS_USER_DOMAIN_ID=default
export OS_PROJECT_NAME=demo
export OS_TENANT_NAME=demo
```

```
export OS_USERNAME=demo
export OS_PASSWORD=password
export OS_AUTH_URL=http://172.29.236.10:5000/v3
```

以降のコマンドは作成した Demo Project に対して実行するため、demo ユーザー用環境変数ファイルを読み込みます。

```
controller_utility_container# source ~/demo-openrc
```

6.4　Glanceへのイメージ登録

cirros のイメージファイルダウンロードし、Glance へ登録します。

イメージファイルのダウンロード

```
controller_utility_container# wget
http://download.cirros-cloud.net/0.3.4/cirros-0.3.4-x86_64-disk.img
--2016-01-25 15:25:52--
http://download.cirros-cloud.net/0.3.4/cirros-0.3.4-x86_64-disk.img
Resolving download.cirros-cloud.net (download.cirros-cloud.net)... 69.163.241.114
Connecting to download.cirros-cloud.net
(download.cirros-cloud.net)|69.163.241.114|:80... connected.
HTTP request sent, awaiting response... 200 OK
Length: 13287936 (13M) [text/plain]
Saving to: 'cirros-0.3.4-x86_64-disk.img'

100%[=================================================>] 13,287,936  2.32MB/s   in 6.3s

2016-01-25 15:25:58 (2.01 MB/s) - 'cirros-0.3.4-x86_64-disk.img' saved
[13287936/13287936]
```

Glance へイメージファイルを登録

```
controller_utility_container# glance image-create --name "cirros-0.3.4-x86_64"
--file cirros-0.3.4-x86_64-disk.img --disk-format qcow2  container-format bare
--progress
[============================>] 100%
+------------------+--------------------------------------+
| Property         | Value                                |
+------------------+--------------------------------------+
| checksum         | ee1eca47dc88f48/9d8a229cc70d07c6      |
| container_format | bare                                 |
| created_at       | 2016-01-25T07:43:37.000000           |
| deleted          | False                                |
```

第 6 章 OpenStack の操作

```
| deleted_at      | None                                 |
| disk_format     | qcow2                                |
| id              | 25751b89-15c5-4903-9135-4a93c66f112a |
| is_public       | False                                |
| min_disk        | 0                                    |
| min_ram         | 0                                    |
| name            | cirros-0.3.4-x86_64                  |
| owner           | 8e19b43b0403413a9f733957d41d4c99     |
| protected       | False                                |
| size            | 13287936                             |
| status          | active                               |
| updated_at      | 2016-01-25T07:43:37.000000           |
| virtual_size    | None                                 |
+-----------------+--------------------------------------+
```

仮想マシンイメージが正しく登録されたか確認します。

```
controller_utility_container# glance image-list
+--------------------------------------+---------------------+-------------+-----------------+----------+--------+
| ID                                   | Name                | Disk Format | Container Format | Size     | Status |
+--------------------------------------+---------------------+-------------+-----------------+----------+--------+
| 25751b89-15c5-4903-9135-4a93c66f112a | cirros-0.3.4-x86_64 | qcow2       | bare            | 13287936 | active |
+--------------------------------------+---------------------+-------------+-----------------+----------+--------+
```

Nova のコマンドラインで Glance と通信して Glance と相互に通信できているかを確認します。

```
controller_utility_container# nova image-list
+--------------------------------------+---------------------+--------+--------+
| ID                                   | Name                | Status | Server |
+--------------------------------------+---------------------+--------+--------+
| 25751b89-15c5-4903-9135-4a93c66f112a | cirros-0.3.4-x86_64 | ACTIVE |        |
+--------------------------------------+---------------------+--------+--------+
```

6.5 仮想ネットワーク設定

続いて、OpenStack 内で利用するネットワークを作成します。ネットワークは外部ネットワークと接続するためのパブリックネットワークと、プロジェクト内で使用するインスタンス用ネットワークを作成します。

パブリックネットワークは既存のネットワークを OpenStack に割り当てます。外部ネットワークへ接続する場合は別途 L3 スイッチ等でネットワークのゲートウェイ IP アドレス、IP アドレスのセグメントを設定しておく必要があります。OpenStack 側に割り当てる IP アドレスの範囲は、そのネットワーク内で DHCP サーバーが動いている場合 DHCP サーバーが配る IP ア

6.5 仮想ネットワーク設定

ドレス範囲を含めないようにしてください。

インスタンスにはインスタンス用ネットワークの範囲の IP アドレスが DHCP Agent を介して割り当てられます。このインスタンスにパブリックネットワークの範囲から Floating IP アドレスを割り当てることで、NAT 接続でインスタンスが外部ネットワークとやり取りができるようになります。

今回はパブリックネットワーク（外部）の IP アドレスは 172.16.100.0/24、インスタンス用ネットワーク（内部）の IP アドレスは 192.168.100.0/24 を使用すると仮定します。

パブリックネットワークの作成

ここでは、neutron コマンドで ext-net と言う名前のパブリックネットワークを作成します。

パブリックネットワークの作成は管理者ユーザーである必要があるため、一度管理者用の環境変数ファイルを読み込みます。

```
controller_utility_container# source ~/openrc
```

パラメータ provider:physical_network で指定する設定は /etc/neutron/plugins/ml2/ml2_conf.ini の physical_interface_mappings に指定されている値を設定します。例えば vlan:eth12 と設定されている場合は vlan と指定します。また、provider:segmentation_id で VLAN ID を指定します。

```
controller_utility_container-acb38a00# neutron net-create  ext-net --router:external
--provider:network_type vlan --provider:physical_network vlan
--provider:segmentation_id 100
Created a new network:
+---------------------------+--------------------------------------+
| Field                     | Value                                |
+---------------------------+--------------------------------------+
| admin_state_up            | True                                 |
| id                        | 7fb8a15f-ba8e-45c0-b513-250c4e2e17e6 |
| mtu                       | 0                                    |
| name                      | ext-net                              |
| provider:network_type     | vlan                                 |
| provider:physical_network | vlan                                 |
| provider:segmentation_id  | 100                                  |
| router:external           | True                                 |
| shared                    | False                                |
| status                    | ACTIVE                               |
| subnets                   |                                      |
| tenant_id                 | b5df55834d9c4b29aa826c3b5b44f2a9     |
+---------------------------+--------------------------------------+
```

パブリックネットワーク用ネットワークアドレス（サブネット）の設定

ext-subnet という名前でパブリックネットワーク用サブネットを作成します。

```
controller_utility_container-acb38a00# neutron subnet-create ext-net 172.16.100.0/24
--name ext-subnet --disable-dhcp --gateway 172.16.100.1
Created a new subnet:
+-------------------+------------------------------------------------------+
| Field             | Value                                                |
+-------------------+------------------------------------------------------+
| allocation_pools  | {"start": "172.16.100.2", "end": "172.16.100.254"}   |
| cidr              | 172.16.100.0/24                                      |
| dns_nameservers   |                                                      |
| enable_dhcp       | False                                                |
| gateway_ip        | 172.16.100.1                                         |
| host_routes       |                                                      |
| id                | b75687f3-53ad-490e-8f77-064e7c24b040                 |
| ip_version        | 4                                                    |
| ipv6_address_mode |                                                      |
| ipv6_ra_mode      |                                                      |
| name              | ext-subnet                                           |
| network_id        | 7fb8a15f-ba8e-45c0-b513-250c4e2e17e6                 |
| subnetpool_id     |                                                      |
| tenant_id         | b5df55834d9c4b29aa826c3b5b44f2a9                     |
+-------------------+------------------------------------------------------+
```

インスタンス用ネットワークの作成

インスタンス用ネットワークを作成するために demo ユーザー用の環境変数を読み込みます。

```
controller_utility_container# source ~/demo-openrc
```

demo-net という名前でインスタンス用ネットワークを作成します。

```
controller_utility_container# neutron net-create demo-net
Created a new network:
+-----------------+--------------------------------------+
| Field           | Value                                |
+-----------------+--------------------------------------+
| admin_state_up  | True                                 |
| id              | a88b6d84-baa6-464e-90a3-b2188b03a2e6 |
| mtu             | 0                                    |
| name            | demo-net                             |
| router:external | False                                |
| shared          | False                                |
| status          | ACTIVE                               |
| subnets         |                                      |
| tenant_id       | 8e19b43b0403413a9f733957d41d4c99     |
+-----------------+--------------------------------------+
```

インスタンス用ネットワークのサブネットを作成

demo-subnet という名前でインスタンス用ネットワークサブネットを作成します。

パラメータ gateway には指定したインスタンス用ネットワークのサブネットの範囲から任意の IP アドレスを指定します。ここでは "192.168.100.0/24" のネットワークをインスタンス用ネットワークとして定義します。

```
controller_utility_container# neutron subnet-create demo-net 192.168.100.0/24 --name demo-subnet
Created a new subnet:
+--------------------+------------------------------------------------------+
| Field              | Value                                                |
+--------------------+------------------------------------------------------+
| allocation_pools   | {"start": "192.168.100.2", "end": "192.168.100.254"} |
| cidr               | 192.168.100.0/24                                     |
| dns_nameservers    |                                                      |
| enable_dhcp        | True                                                 |
| gateway_ip         | 192.168.100.1                                        |
| host_routes        |                                                      |
| id                 | e95ff2a9-7500-4926-8692-5f27a06ca589                 |
| ip_version         | 4                                                    |
| ipv6_address_mode  |                                                      |
| ipv6_ra_mode       |                                                      |
| name               | demo-subnet                                          |
| network_id         | a88b6d84-baa6-464e-90a3-b2188b03a2e6                 |
| subnetpool_id      |                                                      |
| tenant_id          | 8e19b43b0403413a9f733957d41d4c99                     |
+--------------------+------------------------------------------------------+
```

仮想ネットワークルーターの設定

仮想ネットワークルーターを作成して外部接続用ネットワークとインスタンス用ネットワークをルーターに接続し、双方間でデータのやり取りを行えるようにします。

demo-router という名前で仮想ネットワークルーターを作成します。

```
controller_utility_container# neutron router-create demo-router
Created a new router:
+-----------------------+--------------------------------------+
| Field                 | Value                                |
+-----------------------+--------------------------------------+
| admin_state_up        | True                                 |
| external_gateway_info |                                      |
| id                    | 4198471b-1274-47cc-b4e0-723b1ee879cf |
| name                  | demo-router                          |
| routes                |                                      |
| status                | ACTIVE                               |
| tenant_id             | 8e19b43b0403413a9f733957d41d4c99     |
+-----------------------+--------------------------------------+
```

仮想ネットワークルーターとインスタンス用ネットワークを接続します。

```
controller_utility_container# neutron router-interface-add demo-router demo-subnet
```

仮想ネットワークルーターと外部ネットワークを接続します。

```
controller_utility_container# neutron router-gateway-set demo-router ext-net
```

ネットワークの確認

仮想ルーターのゲートウェイ IP アドレスの確認を行います。

neutron router-port-list コマンドを実行すると、仮想ルーターのそれぞれのポートに割り当てられた IP アドレスを確認することができます。コマンドの実行結果から "192.168.100.1" がインスタンスネットワーク側ゲートウェイ IP アドレス、"172.16.100.2" がパブリックネットワーク側ゲートウェイ IP アドレスであることがわかります。

作成したネットワークの確認として、外部 PC からパブリックネットワーク側ゲートウェイ IP アドレスに対して疎通確認を行ってみましょう。問題なければ仮想ルーターと外部ネットワークとの接続ができていると判断することができます。

```
controller_utility_container# neutron router-port-list demo-router -c fixed_ips
--max-width 60
+------------------------------------------------------------+
| fixed_ips                                                  |
+------------------------------------------------------------+
| {"subnet_id": "c7aa9543-8044-4d8d-b087-ceb0968842b5",      |
| "ip_address": "192.168.100.1"}                             |
|                                                            |
+------------------------------------------------------------+
```

作成したポートに対して疎通確認を行います。

```
controller_utility_container# ping -c3 172.16.100.2
PING 172.16.100.2 (172.16.100.2) 56(84) bytes of data.
64 bytes from 172.16.100.2: icmp_seq=1 ttl=62 time=0.401 ms
64 bytes from 172.16.100.2: icmp_seq=2 ttl=62 time=0.383 ms
64 bytes from 172.16.100.2: icmp_seq=3 ttl=62 time=0.389 ms

--- 172.16.100.2 ping statistics ---
3 packets transmitted, 3 received, 0% packet loss, time 1998ms
rtt min/avg/max/mdev = 0.383/0.391/0.401/0.007 ms
```

6.6　セキュリティグループの設定変更

作成したインスタンスが外部からできるように、default ポリシーにセキュリティルールを追加します。

SSH の許可

SSH で使用する TCP プロトコルの 22 番ポートを許可します。

```
controller_utility_container# neutron security-group-rule-create --direction ingress
--ethertype IPv4 --protocol tcp --port-range-min 22 --port-range-max 22 default
```

ICMP エコー要求とエコー応答（ping）の許可

疎通確認 (Ping) で使用する ICMP プロトコルのエコー要求 (Tyep:8) とエコー応答 (Type:0) を許可します。

```
controller_utility_container# neutron security-group-rule-create --direction ingress
--ethertype IPv4 --protocol icmp --port-range-min 8 --port-range-max 8 default

controller_utility_container# neutron security-group-rule-create --direction ingress
--ethertype IPv4 --protocol icmp --port-range-min 0 --port-range-max 0 default
```

以下コマンドでセキュリティルールが追加されたことを確認します。

```
controller_utility_container# neutron security-group-list
+--------------------------------------+---------+------------------------------+
| id                                   | name    | security_group_rules         |
+--------------------------------------+---------+------------------------------+
| fab6c409-acef-424f-b72d-c3c0dabf08dd | default | egress, IPv4                 |
|                                      |         | egress, IPv6                 |
|                                      |         | ingress, IPv4, 22/tcp        |
|                                      |         | ingress, IPv4, icmp (type:8,
code:8)                                 |         |
|                                      |         | ingress, IPv4, icmp (type:0,
code:0)                                 |         |
|                                      |         | ingress, IPv4, remote_group_id:
fab6c409-acef-424f-b72d-c3c0dabf08dd |         |
|                                      |         | ingress, IPv6, remote_group_id:
fab6c409-acef-424f-b72d-c3c0dabf08dd |         |
+--------------------------------------+---------+------------------------------+
```

6.7 インスタンスの起動

プロジェクト内に最低限の構成が準備できたので、実際にインスタンスが起動できるか確認します。

まずは以下の確認コマンドを使ってインスタンスの起動に必要な情報を集めます。

- イメージの一覧表示

```
controller_utility_container# glance image-list
+--------------------------------------+------------------+-------------+------------------+----------+--------+
| ID                                   | Name             | Disk Format | Container Format | Size     | Status |
+--------------------------------------+------------------+-------------+------------------+----------+--------+
| 25751b89-15c5-4903-9135-4a93c66f112a | cirros-0.3.4-x86_64 | qcow2    | bare             | 13287936 | active |
+--------------------------------------+------------------+-------------+------------------+----------+--------+
```

- ネットワークの一覧表示

```
controller_utility_container# neutron net-list
+--------------------------------------+----------+---------------------------------------------------+
| id                                   | name     | subnets                                           |
+--------------------------------------+----------+---------------------------------------------------+
| a88b6d84-baa6-464e-90a3-b2188b03a2e6 | demo-net | e95ff2a9-7500-4926-8692-5f27a06ca589 192.168.100.0/24 |
| 7fb8a15f-ba8e-45c0-b513-250c4e2e17e6 | ext-net  | b75687f3-53ad-490e-8f77-064e7c24b040              |
+--------------------------------------+----------+---------------------------------------------------+
```

- セキュリティグループの一覧表示

```
controller_utility_container# neutron security-group-list
+--------------------------------------+---------+---------------------------------+
| id                                   | name    | security_group_rules            |
+--------------------------------------+---------+---------------------------------+
| fab6c409-acef-424f-b72d-c3c0dabf08dd | default | egress, IPv4                    |
|                                      |         | egress, IPv6                    |
|                                      |         | ingress, IPv4, 22/tcp           |
|                                      |         | ingress, IPv4, icmp (type:8, code:8) |
```

6.7 インスタンスの起動

```
|                                              |         | ingress, IPv4, remote_group_id: 
 fab6c409-acef-424f-b72d-c3c0dabf08dd |
|                                              |         | ingress, IPv6, remote_group_id: 
 fab6c409-acef-424f-b72d-c3c0dabf08dd |
+----------------------------------------------+---------+-------------------------------
```

- フレーバーの一覧表示

```
controller_utility_container# nova flavor-list
+----+-----------+-----------+------+-----------+------+-------+-------------+-
| ID | Name      | Memory_MB | Disk | Ephemeral | Swap | VCPUs | RXTX_Factor | Is_Public |
+----+-----------+-----------+------+-----------+------+-------+-------------+-
| 1  | m1.tiny   | 512       | 1    | 0         |      | 1     | 1.0         | True |
| 2  | m1.small  | 2048      | 20   | 0         |      | 1     | 1.0         | True |
| 3  | m1.medium | 4096      | 40   | 0         |      | 2     | 1.0         | True |
| 4  | m1.large  | 8192      | 80   | 0         |      | 4     | 1.0         | True |
| 5  | m1.xlarge | 16384     | 160  | 0         |      | 8     | 1.0         | True |
+----+-----------+-----------+------+-----------+------+-------+-------------+-
```

集めた情報を元に nova boot コマンドを使ってインスタンスを起動します。インスタンスを起動するために必要な項目は以下のとおりです。

項目	値
インスタンス名	vm1
イメージ	cirros-0.3.4-x86_64
フレーバー	m1.tiny
ネットワーク	demo-net（コマンドでは ID を指定）
セキュリティグループ	default（コマンドでは ID を指定）

- インスタンスの起動

```
controller_utility_container# nova boot --flavor m1.tiny --image 
"cirros-0.3.4-x86_64" --nic net-id=a88b6d84-baa6-464e-90a3-b2188b03a2e6 
--security-group=fab6c409-acef-424f-b72d-c3c0dabf08dd vm1

+--------------------------------------+---------------------------------
| Property                             | Value
+--------------------------------------+---------------------------------
| OS-DCF:diskConfig                    | MANUAL
```

```
| OS-EXT-AZ:availability_zone          | nova                                 |
| OS-EXT-STS:power_state               | 0                                    |
| OS-EXT-STS:task_state                | scheduling                           |
| OS-EXT-STS:vm_state                  | building                             |
| OS-SRV-USG:launched_at               | -                                    |
| OS-SRV-USG:terminated_at             | -                                    |
| accessIPv4                           |                                      |
| accessIPv6                           |                                      |
| adminPass                            | hRiWzPNwvp8B                         |
| config_drive                         |                                      |
| created                              | 2016-01-25T07:47:28Z                 |
| flavor                               | m1.tiny (1)                          |
| hostId                               |                                      |
| id                                   | 05477352-6652-4087-9a93-7376e327d5cb |
| image                                | cirros-0.3.4-x86_64                  |
  (25751b89-15c5-4903-9135-4a93c66f112a) |
| key_name                             | -                                    |
| metadata                             | {}                                   |
| name                                 | vm1                                  |
| os-extended-volumes:volumes_attached | []                                   |
| progress                             | 0                                    |
| security_groups                      | 835a57d5-37b0-4e4a-a421-76547d527f3d |
| status                               | BUILD                                |
| tenant_id                            | 8e19b43b0403413a9f733957d41d4c99     |
| updated                              | 2016-01-25T07:47:29Z                 |
| user_id                              | 776a28828ff44378bc5a4efa26c36fc6     |
+--------------------------------------+--------------------------------------+
```

インスタンス起動後正しく起動できたかコンソールログを確認します。

```
controller_utility_container# nova console-log 05477352-6652-4087-9a93-7376e327d5cb
```

6.8 FloatingIPの設定

外部からインスタンスへのアクセスを可能とするために以下コマンドでFloatingIPの設定を行います。

インスタンス一覧を表示し起動したインスタンスのIDを確認します。

```
controller_utility_container# nova list
+--------------------------------------+------+--------+------------+-------------+
| ID                                   | Name | Status | Task State | Power State |
Networks                       |
+--------------------------------------+------+--------+------------+-------------+
| 05477352-6652-4087-9a93-7376e327d5cb | vm1  | ACTIVE | -          | Running     |
demo-net=192.168.100.3 |
+--------------------------------------+------+--------+------------+-------------+
```

Neutronのポート一覧より該当インスタンスで使用しているポートIDを検索します。

```
controller_utility_container# neutron port-list | grep 192.168.100.3
| c044267e-0902-4fff-9909-1b05d9204683 |        | fa:16:3e:7d:49:92 | {"subnet_id":
"e95ff2a9-7500-4926-8692-5f27a06ca589", "ip_address": "192.168.100.3"} |
```

FloatingIPを作成します。

```
root@controller-01_utility_container# neutron floatingip-create ext-net
Created a new floatingip:
+---------------------+--------------------------------------+
| Field               | Value                                |
+---------------------+--------------------------------------+
| fixed_ip_address    |                                      |
| floating_ip_address | 172.16.100.6                         |
| floating_network_id | 8ee3a91d-a516-4962-bbe4-ab8ecfcd92b2 |
| id                  | ab123966-87a1-401a-8c9c-b5b818a14d1b |
| port_id             |                                      |
| router_id           |                                      |
| status              | DOWN                                 |
| tenant_id           | 81ce71826dd24dc7a39d59b54b9c21bf     |
+---------------------+--------------------------------------+
```

FloatingIPの一覧を表示し、正しく作成されたかどうか確認します。

構文

第6章 OpenStack の操作

```
neutron floatingip-list [FloatingIP の ID] [ポート ID]
```

実行例

```
controller_utility_container# neutron floatingip-list
ab123966-87a1-401a-8c9c-b5b818a14d1b daf892fa-1485-4568-a15a-70483b5bd1a7
+--------------------------------------+------------------+---------------------+
| id                                   | fixed_ip_address | floating_ip_address |
port_id |
+--------------------------------------+------------------+---------------------+
| ab123966-87a1-401a-8c9c-b5b818a14d1b |                  | 172.16.100.3        |
|
+--------------------------------------+------------------+---------------------+
```

検索したポートへ FloatingIP を割り当てます。

```
controller_utility_container# neutron floatingip-associate
4e5e5f46-a08e-4764-920e-a2a78a741cb1 c044267e-0902-4fff-9909-1b05d9204683
Associated floating IP ab123966-87a1-401a-8c9c-b5b818a14d1b
```

もう一度インスタンス一覧を表示し、FloatingIP が割り当てられたことを確認します。

```
controller_utility_container# nova list
+--------------------------------------+------+--------+------------+------------+
| ID                                   | Name | Status | Task State | Power State |
Networks                              |
+--------------------------------------+------+--------+------------+------------+
| 05477352-6652-4087-9a93-7376e327d5cb | vm1  | ACTIVE | -          | Running    |
demo-net=192.168.100.3, 172.16.100.3 |
+--------------------------------------+------+--------+------------+------------+
```

6.9　Cinder の設定

以下コマンドで Cinder ボリュームを作成し、正常に Cinder が動作していることを確認します。以下の例では volume01 という名前でサイズ 2GB のボリュームを作成します。

```
controller_utility_container# cinder create --display-name volume01 2
```

ボリュームが作成されていることを確認します。

```
controller_utility_container# cinder list
+--------------------------------------+-----------+--------------+------+-------
|                  ID                  |  Status   | Display Name | Size | Volume
Type | Bootable | Attached to |
+--------------------------------------+-----------+--------------+------+-------
| e28b625e-2239-4d89-a809-a0d98abaa505 | available |   volume01   |  2   | None
| false    |             |
```

```
+----------+-------------------------------------+-----------+---------------+------+--------
```

作成したボリュームをインスタンスに割り当てます。

```
controller_utility_container# nova volume-attach vm1
e28b625e-2239-4d89-a809-a0d98abaa505
+----------+--------------------------------------+
| Property | Value                                |
+----------+--------------------------------------+
| device   | /dev/vdb                             |
| id       | e28b625e-2239-4d89-a809-a0d98abaa505 |
| serverId | 05477352-6652-4087-9a93-7376e327d5cb |
| volumeId | e28b625e-2239-4d89-a809-a0d98abaa505 |
+----------+--------------------------------------+
```

再びボリューム一覧を表示して、正しく割り当てられたことを確認します。

```
controller_utility_container# cinder list
+--------------------------------------+--------+--------------+------+-------------+
|                  ID                  | Status | Display Name | Size | Volume Type |
Bootable |       Attached to           |
+--------------------------------------+--------+--------------+------+-------------+
| e28b625e-2239-4d89-a809-a0d98abaa505 | in-use |   volume01   |  2   |    None     |
|  false  | 05477352-6652-4087-9a93-7376e327d5cb |
+--------------------------------------+--------+--------------+------+-------------+
```

6.10　インスタンスの動作確認

最後に、作成したインスタンスへ SSH でログインし正しく動作するか確認します。

- インスタンスへの疎通確認

```
ping [割り当てた FloatingIP アドレス]
```

- SSH ログイン

```
ssh cirros@[割り当てた FloatingIP アドレス]
cirros@172.16.100.3's password:   ← パスワードは cubswin:)
```

- 割り当てた Cinder ボリュームにパーティションを作成

第 6 章　OpenStack の操作

```
$ sudo fdisk /dev/vdb

Command (m for help): n

Partition type:
   p   primary (0 primary, 0 extended, 4 free)
   e   extended
Select (default p): p

Partition number (1-4, default 1): ← Enter
Using default value 1

First sector (2048-4194303, default 2048): ← Enter
Using default value 2048

Last sector, +sectors or +size{K,M,G} (2048-4194303, default 4194303): ← Enter
Using default value 4194303

Command (m for help): w

The partition table has been altered!
Calling ioctl() to re-read partition table.
Syncing disks.
```

- 作成したパーティションをフォーマット

```
$ sudo mkfs /dev/vdb1

mke2fs 1.42.2 (27-Mar-2012)
Filesystem label=
OS type: Linux
Block size=4096 (log=2)
Fragment size=4096 (log=2)
Stride=0 blocks, Stripe width=0 blocks
131072 inodes, 524032 blocks
26201 blocks (5.00%) reserved for the super user
First data block=0
Maximum filesystem blocks=536870912
16 block groups
32768 blocks per group, 32768 fragments per group
8192 inodes per group
Superblock backups stored on blocks:
    32768, 98304, 163840, 229376, 294912

Allocating group tables: done
Writing inode tables: done
Writing superblocks and filesystem accounting information: done
```

- 作成したパーティションをマウント

6.10 インスタンスの動作確認

```
$ sudo mount /dev/vdb1 /mnt
```

- パーティションのサイズ確認

/mnt へ Cinder ボリュームがマウントされていることを確認します

```
$ df -h
Filesystem               Size     Used Available Use% Mounted on
/dev                    242.3M       0    242.3M   0% /dev
/dev/vda1                23.2M   18.0M      4.0M  82% /
tmpfs                   245.8M       0    245.8M   0% /dev/shm
tmpfs                   200.0K   88.0K    112.0K  44% /run
/dev/vdb1                 2.0G    3.0M      1.9G   0% /mnt
```

第7章 Tipsとトラブルシューティング

この章では、OpenStack-Anibleに関連するトラブルの回避方法について説明します。

7.1 コンテナを削除し再構築する

OpenStack-Ansibleでは、環境を一から構築しなおしたいときのために既存の環境を削除するスクリプトが用意されています。

このスクリプトを実行すると/etc/openstack_deploy配下で設定したファイルは残りますが、**デプロイ実施後に構築されたもの（コンテナやDB等）がすべて削除**されます。必要に応じて実行前にバックアップを取っておいてください。

```
controller# cd /opt/openstack-ansible/
controller# scripts/teardown.sh
------------------------------------------------------------------------
WARNING: This is a destructive action. All containers will be destroyed and
         all data within the containers will be removed.  Some data will be
         removed from the host as well. /etc/openstack_deploy will be preserved
         and may be manually removed if needed.
         Please verify that you have backed up all important data prior to
         proceeding with the teardown script.
------------------------------------------------------------------------
To REALLY destroy all containers and delete the data within them,
type 'Y' or 'y' and press enter:
```

7.2 OpenStack 構築後 Controller サーバーを再起動すると正常に動作しない

Controller サーバーを再起動するとコンテナの起動順等の理由により Galera クラスタ (DB サービス) が正しく起動できないことがあります。この場合、以下の手順で Galera クラスタのリカバリーを実行してください。

```
controller# cd /opt/openstack_deploy/playbook
controller# openstack-ansible -e galera_ignore_cluster_state=true galera-install.yml
```

7.3 ネットワークの冗長化

Linux では Bonding モジュールを利用することで、複数の NIC を束ねネットワークの冗長化や負荷分散を行うことができます。

https://help.ubuntu.com/community/UbuntuBonding

OpenStack-Ansible で構築された OpenStack 環境はほぼコンテナ上で稼働しているため、サーバーの物理 NIC で Bonding を組むことにより比較的容易にネットワークの冗長化が実現できます。OpenStack Foundation の公式マニュアルでは /etc/network/interfaces ファイルの設定例が紹介されています。詳しくは以下のページをご確認ください。

http://docs.openstack.org/developer/openstack-ansible/install-guide/targethosts-networkrefarch.html

7.4 YAML ファイルの構文について

OpenStack-Ansible では、設定ファイルに YAML 形式を使用しています。YAML 形式ではインデントを使って階層構造を組むため、カラム位置が重要になっています。ファイルを編集するときは以下の点に注意してください。

- YAML でサブ項目を複数組むときはカラム位置がずれないよう気をつけてください。一般的には半角スペース 2 文字でインデントを空けます
- インデントで「タブ文字」を使用することはできません
- コメント行は「#」を使用します

YAML ファイル 構文の例

```
---
shared-infra_hosts:
  controller-01:
#  ↓ タブ文字は使用できない
    affinity:
      galera_container: 3
      rabbit_mq_container: 3
    ip: 172.29.236.10
#    ↑ 位置のズレに注意
```

YAML ファイルの構文チェック (yamllint)

　yamllint を使用することで作成した YAML ファイルの構文に誤りがないかチェックすることができます。yamalint のインストール方法は以下のとおりです。

```
$ sudo pip install yamllint
```

　yamllint の使用方法は以下のとおりです。

```
### YAML ファイルを個別に指定する
$ yamllint file1.yml file2.yml ...

### ディレクトリ内の YAML ファイルをすべて指定する
$ yamllint .
```

付録A　FAQフォーラム参加特典について

　本書の購入者限定特典としてFAQフォーラム（Googleグループ）を用意しています。GoogleアカウントにログインのうえT記URLにアクセスし、本フォーラムの「メンバー登録を申し込む」をクリックしてください。申し込みの際に追加情報として購入した書籍名をご入力ください。

https://groups.google.com/d/forum/vtj-openstack-faq

　本フォーラムは書籍の購入者限定のサービスです。フォーラム内のすべての質問に対して日本仮想化技術の社員が回答を行うことをお約束するものではないこと、サービスの公開期間は書籍出版後2年間（〜2018/03）を予定していること、前記期間内であってもOpenStackおよび関連サービスの仕様変更に伴いフォーラムを継続できなくなる可能性があることをご了承ください。また、本書およびフォーラムの解説内容によって生じる直接または間接被害について、著者である日本仮想化技術株式会社ならびに株式会社インプレスでは一切の責任を負いかねます。

●著者紹介

田口 貴久
日本仮想化技術株式会社
中堅システム会社にて基幹系システムの運用 SE 等の経て、2012 年より日本仮想化技術株式会社に所属。現在は仮想化インフラの構築・運用や OpenStack デプロイツールの調査・検証等の担当。長年の運用経験を生かし、よりよいインフラ環境の構築を目指して日々奮闘中。

目次 英人
NEC ネッツエスアイ株式会社
NEC ネッツエスアイ株式会社 パブリックソリューション事業部に所属。日本仮想化技術株式会社と協業し、OpenStack を中心としたクラウド基盤の構築・運用のノウハウを蓄積中。

●スタッフ
- 田中 佑佳（表紙デザイン）
- 鈴木 教之（編集、紙面レイアウト）

本書のご感想をぜひお寄せください
http://book.impress.co.jp/books/1115101151
アンケート回答者の中から、抽選で商品券（1万円分）や図書カード（1,000円分）などを毎月プレゼント。
当選は商品の発送をもって代えさせていただきます。

●本書の内容に関するご質問は、書名・ISBN・お名前・電話番号と、該当するページや具体的な質問内容、お使いの動作環境などを明記のうえ、インプレスカスタマーセンターまでメールまたは封書にてお問い合わせください。電話やFAX等でのご質問には対応しておりません。なお、本書の範囲を超える質問に関しましてはお答えできませんのでご了承ください。

●落丁・乱丁本はお手数ですがインプレスカスタマーセンターまでお送りください。送料弊社負担にてお取り替えさせていただきます。但し、古書店で購入されたものについてはお取り替えできません。

■読者の窓口
インプレスカスタマーセンター
〒101-0051 東京都千代田区神田神保町一丁目105番地
　TEL　03-6837-5016　／　FAX　03-6837-5023
info@impress.co.jp

■書店／販売店のご注文窓口
株式会社インプレス 受注センター
　TEL　048-449-8040
　FAX　048-449-8041

OpenStack-Ansibleで作る
HA環境構築手順書 Kilo版（Think IT Books）

2016年5月1日 初版発行

著　者　日本仮想化技術株式会社
発行人　土田 米一
編集人　高橋 隆志
発行所　株式会社インプレス
　　　　〒101-0051　東京都千代田区神田神保町一丁目105番地
　　　　TEL　03-6837-4635（出版営業統括部）
　　　　ホームページ　http://book.impress.co.jp/

本書は著作権法上の保護を受けています。本書の一部あるいは全部について（ソフトウェア及びプログラムを含む）、株式会社インプレスから文書による許諾を得ずに、いかなる方法においても無断で複写、複製することは禁じられています。

Copyright © 2016 Virtual Tech JP. All rights reserved.
印刷所　京葉流通倉庫株式会社
ISBN978-4-8443-8058-0　C3055
Printed in Japan